Education
141

化装舞会

A Dress-up Party

Gunter Pauli

[比] 冈特·鲍利　著

[哥伦] 凯瑟琳娜·巴赫　绘

何家振　译

上海远东出版社

丛书编委会

主　任：田成川

副主任：闫世东　林　玉

委　员：李原原　祝真旭　曾红鹰　靳增江　史国鹏
　　　　梁雅丽　孟小红　郑循如　陈　卫　任泽林
　　　　薛　梅　朱智翔　柳志清　冯　缨　齐晓江
　　　　朱习文　毕春萍　彭　勇

特别感谢以下热心人士对童书工作的支持：

匡志强　宋小华　解　东　厉　云　李　婧　庞英元
李　阳　梁婧婧　刘　丹　冯家宝　熊彩虹　罗淑怡
旷　婉　王靖雯　廖清州　王怡然　王　征　邵　杰
陈强林　陈　果　罗　佳　闫　艳　谢　露　张修博
陈梦竹　刘　灿　李　丹　郭　雯　戴　虹

目录

Contents

ZERI Learning Initiative

青蛙遇到了一只水蚤，他想起来曾经在池塘里和池塘周围看见过她和很多其他水蚤。

"很高兴又见到你。"青蛙向漂亮的水蚤打招呼。

"你是谁？我从来没有见过你。如果你能离我远一点，我将不胜感激。"

A frog meets a nymph and remembers seeing her and many others in and around the pond.

"Good to see you again," the frog greets the pretty nymph.

"Who are you? I have never met you before. So I would appreciate if you keep your distance."

青蛙遇到了一只水虿......

A frog meets a nymph ...

你不记得我吗?

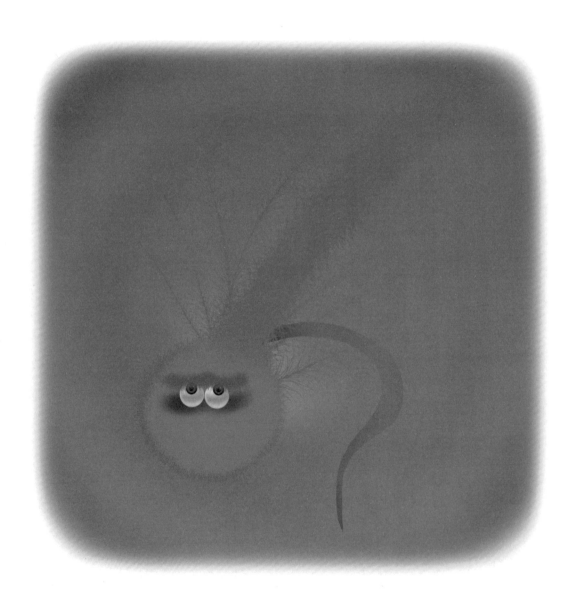

you don't remember me?

"你不记得我吗？我一直在这里游泳，在这个池塘里好几个星期了。来，再看看我。"

"我的记忆力很好，"水蚤回答道，"但我从来没在这里见过你这样的。"

"你是睡眠不足，还是缺少维生素呢？"

"You don't remember me? I have been swimming here, in this pond, for weeks. Come on, take another look."

"I have a very good memory," the nymph responds, "but you do not look like anything I have ever seen here before."

"Do you not get enough sleep, or are you perhaps lacking an important vitamin?"

"听着，先生。"她回答道，"你想方设法与我攀谈。我告诉你，我绝对绝对没见过你。所以，你能别打扰我了吗？"

"但是，是我呀！你的朋友，那只青蛙呀。"

"Look here, mister," she replies. "You have tried every trick in the book to try and strike up a conversation with me. I tell you, I have never-ever seen you before. So would you please leave me alone?"

"But it is me! Your friend, the frog."

但是，是我呀！你的朋友，那只青蛙呀。

But it is me! Your friend, the frog.

你甚至让我想起了恐龙。

you even make me think of the dinosaurs.

"最近我们在这里唯一的朋友是蝌蚪。我们看着他们吃植物和藻类，而你看起来是典型的食肉动物。你甚至让我想起了恐龙。"

"这么说你的确认识那些蝌蚪，对吧？我曾经就是他们中的一员！"

"The only friends we have had around here lately were the tadpoles. We watched them eat plants and algae, but you have the typical face of a carnivore. You even make me think of the dinosaurs."

"So you do remember the tadpoles? I was one of them!"

"那又怎样？你假装是一只蝌蚪，一只只有前腿，有胡须、鳃片和螺旋形嘴的小蝌蚪，然后一夜之间变成了有强壮的后腿、巨大的颌骨和长长的舌头的动物，准备好了吃我。"

"但是，我就是那只蝌蚪哇！"

"算了吧，我以前也去过化装舞会，很多人外表变化惊人，但我可从没见过谁的外表变化像你这样大的。"

"How about that? You pretend that a tadpole, one with only front legs, who has whiskers and fins, and a spiral-shaped mouth is transformed overnight into an animal with strong back legs, a big jaw and a long tongue – ready to eat me."

"But it is me!"

"Well, I have been to dress up parties before, ones with many surprises, but never-ever have I seen anyone change his appearance like this!"

你假装是一只蝌蚪!

You pretend you were a tadpole!

所有的青蛙一开始都是蝌蚪……

All frogs are tadpoles first ...

"但是，这是真的。所有的青蛙一开始都是蝌蚪，然后他们就变成了……"

"食肉动物。当你是蝌蚪时，你主要吃植物，但当你变成青蛙后，你就会想吃我。"

"好吧，我的小水蚤朋友，将来你的外表也会变化，可能比我的变化还大。几个星期后，你会变成什么呢？"

"But it is true. All frogs are tadpoles first and then they become ..."

"Carnivores. When you are a tadpole you eat mostly plants, when you are a frog you will want to eat me."

"Well, my little nymph friend, you will change your looks as well. Perhaps even more than I do. What will you be in a couple of weeks?"

"我会变成一只蜻蜓。"她骄傲地回答。

"你？你会变成一种曾在地球上生活过，甚至在恐龙在地球上漫步之前就到处飞翔的史前昆虫？"

"没错！但是作为水虿，我住在水里。"

"我知道。我以前常和你一起玩呢。"

"I will be a dragonfly," she says proudly.

"You? You will turn into one of those prehistoric insects that lived on earth, flying around everywhere, from even before the dinosaurs roamed the Earth?"

"Exactly! But as a nymph I live in water."

"I know. I used to play with you."

我会变成一只蜻蜓。

I will be a dragonfly.

······我用屁股呼吸。

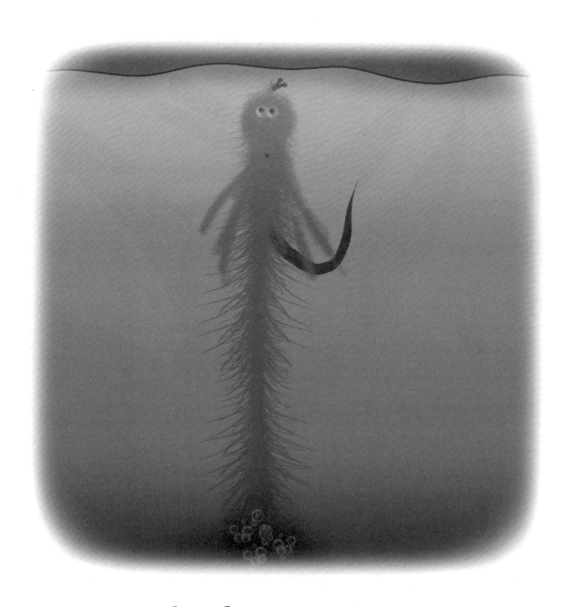

... I breathe through my butt.

"当我准备离开水面时，我就会蜕皮，变成一只蜻蜓。"

"我一直很好奇，你怎么能在水中游动呢。你没有翅膀，没有鳃，也没有腿。"

"那是因为我用屁股呼吸。"

"你开玩笑吧？没人用屁股呼吸，更别说利用屁股呼吸推动自己前进了。"

"When I am ready to get out of the water I will shed my skin and turn into the dragonfly."

"I always wondered how you are able to move around in the water. You don't have wings, or fins or legs."

"That is because I breathe through my butt."

"You are joking! No one breathes through their backside, let alone making use of that to propel themselves forward."

"嘿，小青蛙，你要学习的东西还多着呢。我告诉你，我把水吸进屁股，提取出我需要呼吸的空气，然后排出水，这样就能推动我前进。"

"既然这样，如果你想在下一次化装舞会上款待我的朋友们，你可以演示一下你怎么做到的——我们肯定会笑死的！"

……这仅仅是开始！……

"Look, you are a young frog and still have to learn a lot. I tell you, I draw water into my butt, take out the air I need to breathe, and then expel the water. That pushes me forward."

"Now, if you would like to entertain my friends at our next dress-up party, you can show us how you do that – and we will have the greatest laugh ever!"

... AND IT HAS ONLY JUST BEGUN! ...

......这仅仅是开始! ......

... AND IT HAS ONLY JUST BEGUN! ...

Did You Know?

你知道吗？

In Greek mythology a nymph is a natural spirit that looks like a beautiful maiden, living in rivers or woods. In nature it is an immature form of a dragonfly.

在希腊神话中，宁芙（水蚤）是居于山林水泽的仙女，而在自然界中，水蚤是蜻蜓的幼虫。

3 年

The dragonfly will spend 90%-95% of its life as a nymph, a development that can take from a few weeks to up to 3 years. Dragonfly larvae may moult 6 to 15 times.

蜻蜓一生中 90%—95% 的时间是以水蚤的形态度过的，这个阶段短则几周，长的可达 3 年。水蚤可能会经历 6 至 15 次蜕皮才能成为蜻蜓。

蜻蜓经历了一个不完全变态过程，跳过了蛹阶段。水虿需要花3个小时从水中爬出变成蜻蜓，在这个过程中，它丢掉了口器，向翅膀框架注入液体，待在阳光下晒干获得翅膀。

Dragonflies undergo an incomplete metamorphosis, skipping the pupal stage. It takes the nymph three hours to turn into a dragonfly by crawling out of the water, losing its mouth part, gaining wings by pumping liquid into their frames, and sitting in the sun to dry.

There are nearly 4,000 species of frogs. There are 5,000 types of dragonflies. It is difficult to distinguish them in their tadpole or nymph stage.

青蛙有将近 4 000 种。蜻蜓约有 5 000 种。很难区分蝌蚪和水虿。

一些青蛙将卵产在池塘或
小溪边的树枝上，它们的卵看
起来像一团泡沫状的茧。下雨
时，小蝌蚪从泡沫里滴到水
中。看起来像是在下蝌蚪雨。

Some frogs lay their eggs in a foamy mass cocoon on a tree branch hanging over a pond or a stream. When it rains, the foam drips down dropping tiny tadpoles into the water below. It may seem like it is raining tadpoles.

刚孵出时，蝌蚪以其内脏
里的卵黄为食，之后，它就以
藻类为食。蝌蚪会长出细小的
像牙齿一样的褶皱将食物摩擦
成小颗粒，帮助进食。一旦蝌
蚪长出腿来，它们的食谱中就
包括昆虫了。

After hatching, the tadpole at first feeds off the yolk in its guts, and then it will feed on algae. Tadpoles will develop tiny tooth-like ridges that help grate food, plants in particular. Once they have legs, their diet includes insects.

Frogs' tongues are attached to the front of their mouths rather than at the back, like those of humans. A frog catches an insect by casting its sticky tongue from its mouth and wraps it around the prey.

青蛙的舌头不像人类的舌头在嘴的后部，而是贴近嘴的前部，青蛙从嘴里吐出具有黏性的舌头，然后将它的猎物卷回来。

Frogs have a moist and slimy skin. Toads have a dry, bumpy skin that tastes bitter to animals trying to catch it. Their skin produces an odour that smells like a skunk. Toads can survive far from water. Toads run or make small hopping movements, while frogs take long, high leaps.

青蛙的皮肤潮湿而黏滑。蟾蜍的皮肤干燥且崎岖不平，让那些想捕食它们的动物感到苦涩。蟾蜍的皮肤产生一种类似臭鼬的气味，它们可以在远离水的地方生存。蟾蜍能够跑动或者小幅跳跃，而青蛙跳得高而远。

Think About It

想一想

Would you recognise every one of your friends at a dress-up party?

你能在化装舞会上认出每一个朋友吗?

How would you respond if you recognise someone, but that person does not recognise you?

如果你认出一个朋友,他却没认出你,你会有什么反应?

Who will be happy to have a chance, not just to dress up, but to transform your looks forever – with no way to return to your old look?

如果有机会永久改变外表,并且无法回到原来的模样（而不仅仅是化装）,谁会为此感到高兴呢?

Do you like surprises? Or do you prefer that before things change you are quietly and carefully informed so as not to receive a shock?

你喜欢惊喜吗? 还是喜欢在事情变化前,你就悄无声息地谨慎得知, 以免感到震惊?

Dress up time! Find something to wear that your siblings and family have never seen. Make sure that it includes a wig and use make up that changes the colour of your face to one that no one associates with you. Also learn how to change your voice. How easy is it for you to confuse those who know you well, and make it impossible for them to recognise you? How successful are you at this? What do you think are the most effective ways to achieve this? Share your results with your friends.

化装时间到了！找一些你父母和兄弟姐妹从来没见过的东西穿上，一定要有一副假发，通过化妆改变脸色，确保没人会想到是你。也要学会怎样改变你的声音。看看能迷惑哪些非常熟悉你的人，让他们认不出来有多容易？你成功的可能性有多大？你认为什么方法最有效呢？与朋友们分享你的结论。

学科知识

Academic Knowledge

生物学	青蛙是有脊椎冷血两栖动物；缺乏营养素会导致失忆，特别是缺乏维生素B$_{12}$；蜻蜓的眼睛占据了它头部的大部分区域；成年蜻蜓每日吃30至100只蚊子，是抑制蚊子种群的重要力量；为适应水中生存，水蚤具有从水中分离空气和氧的能力。
化 学	胶原酶，一种降解胶原蛋白的酶，在脊椎动物、无脊椎动物和植物中广泛存在；在再生过程中一向以正效应而著称的抗氧化剂，在癌症治疗之后的组织再生过程中，却似乎具有负效应。
物 理	在类风湿性关节炎中软骨被破坏，减震性能消失，使病人的关节损坏并且疼痛；蜜蜂翅膀每秒钟振动约200—240次，蜻蜓翅膀只振动约30次；蜻蜓的眼睛有3万个复眼；青蛙的喉咙里有一个装满空气的小囊，囊中空气慢慢释放，发出了哇哇的叫声。
工程学	在蝌蚪尾巴里发现的软骨组织，含有很多蛋白聚糖，能起到减震作用；蜻蜓能直上直下地飞，并且能像直升机那样盘旋；蜻蜓用它们的脚捕捉猎物，命中率达到90%以上。
经济学	为了取得长远的成功，企业必须进行重大变革，而且这种变革可以是破坏性的（约瑟夫·熊彼特），迫使公司里很多人适应新的环境，学习新的专业，调整自我，这就是如同蝌蚪变成青蛙或水蚤变成蜻蜓一样的根本的变革。
伦理学	生活面临不断变化的挑战；人们通常不愿意或者害怕改变，并试图强迫一切事物都保持现状，但是为了生存和发展，我们必须适应调整和转变。
历 史	蜻蜓大概是第一种有翼昆虫，它们大约存在了3亿年；历史上，蜻蜓象征成熟和深沉的性格。
地 理	蜻蜓可以来回跨越印度洋，在所有昆虫中迁徙距离最长；"变形"来自希腊语，意思是"改变"。
数 学	猎物判断蜻蜓是不是静止的唯一方法是看捕食者接近它时体型的大小变化，这个问题是摄影几何学的一部分，摄影几何学是在几何透视研究历史中发展起来的一门数学学科。
生活方式	万圣节前夕是世界各地孩子们最喜欢的化装时间；使用一种流行的玩具"伪装机器人"（也被称为变形金刚），人可以变成机器。
社会学	温水煮青蛙的故事（查尔斯·汉迪）讲的是，尽管一个人面临生命威胁，必须要采取主动行动来改变现状，但却无动于衷，故步自封；面对不同的情况，我们需要适应不同的社会角色。
心理学	人从幼儿阶段起就喜欢穿各种服装：摆脱日常生活变成一个牛仔、公主或者老虎；因为穿着变化，自我感觉和行为习惯也发生改变。
系统论	在生活中变革是必需的，然而，人们倾向宁可改变周围的事物，而不是使自己适应新的环境，如年龄增长和职业改变。

情感智慧
Emotional Intelligence

青蛙

青蛙是池塘里快乐的居民。他很执着，哪怕是在水蚤不仅不认识他，而且拒绝与他交谈的时候。他很自信，多次被拒绝却不受影响，坚持追问水蚤不认识他的原因，以至于让水蚤感受到了威胁。青蛙寻找唤起水蚤美好回忆的方法，两次坚持说他真的就是从前那只蝌蚪。当被指控为食肉动物时，青蛙既没有承认也没有驳斥，而是通过问水蚤变形成为蜻蜓的问题转移话题。当得知水蚤如何在水中呼吸和游动后，青蛙从吃惊转化为开玩笑。

水蚤

水蚤希望青蛙与她保持距离，因为水蚤将青蛙视为安全威胁。她坚持要独自待着，并且要求青蛙不要装作与自己相识并试图进行交谈。水蚤清楚青蛙是食肉动物，很危险，有可能将自己变成一顿美餐。她仔细分析了青蛙的变形，在了解了青蛙从蝌蚪转化成青蛙的过程之后，并没有显示出一点热情。当青蛙指出她将从水蚤变成蜻蜓之后，她依然无动于衷。水蚤不带感情、不谈细节地说明她的想法。水蚤并不认为自己能用屁股呼吸很特别，水蚤要求青蛙学习更多大自然的知识；当青蛙认为这个能力非常滑稽时，水蚤保持了她原有的方式，不动声色。

艺术
The Arts

不同的青蛙叫声区别很大，听起来可能像"唧唧"声、口哨声、"呱呱"声、"哞哞"声、"吱吱"声、"咯咯"声、狗叫声和"咕哝"声。现在听一听这些不同的声音，试着模仿一下。当几种不同的蛙叫声同时响起时，听起来像什么？这样听起来会很和谐，还是会很刺耳呢？

思维拓展
Systems: Making the Connections

我们周围的事物总是在变化的——以至于我们的认识无法跟上那些正在发生的变化。挑战在于人们喜欢维持现状，做出极大努力并花费大额资金，让所有事情都保持原状——更有甚者，把事情弄得比原先看起来更好，比如对身体进行整形手术使衰老的外表看起来年轻，或者造出理想的容貌。自然界展示了令人困惑的调整、改变、转化、适应甚至变形的能力。自然界的很多物种知道在必要时如何隐藏和伪装。通过这些方式，它们以不同的形式欢度它们的生活，这些改变和变形策略是它们维持持续适应能力的一部分，是确保它们获得更多水和食物的有效方式，并且维持物种在生态系统中已获取的位置。人类需要认识到，我们是周围正在发生变化的原因，虽然我们应该对环境、生态和气候变化负部分责任，但是我们却拒绝改变我们自己。我们需要努力避免摧毁人类赖以生存的条件。第一步是理解我们周围每件事物是如何运行的，虽然一些科学发现令人惊奇，甚至看起来奇怪或好笑，人类应该深刻理解我们周围生命系统用于应对变化独具匠心的方法。它们利用温度、湿度、气压、食物可得性等外部变化，追求进化和共生的路径，最终实现生命的华章。如果人类要生存，那么我们就要改变我们的生活、生产和消费方式。

动手能力
Capacity to Implement

列出自然界最惊人的十大变形。你知道了蝌蚪和水蚤的故事，但是还有更多。我们不应该只看物理变化，还要看生物物种用化学方法表演的惊人的特技，比如，大马哈鱼必须使其能够从在淡水中生活，转到海水中生活，然后再回到淡水中生活。人也是不错的变形者：出生于纯液体的细胞，在体内生长出骨骼，当年龄增长时，骨头长得更长、更粗。

故事灵感来自
This Fable Is Inspired by

玛丽·施韦策
Mary Schweitzer

　　玛丽·施韦策是一位古生物学家，她起初在犹他大学主修传播学，后来在蒙大拿大学获得第二教育证书，并获得生物学博士学位。她是第一个从恐龙化石中辨认和分离软组织（胶原）的研究者，她注意到在实验室降解的化学品，在受保护的自然环境中并不降解。施韦策还发现铁粒子在保存软组织的过程中可能起一定作用。她非常热衷于弄懂蝌蚪的尾巴是如何转变的，以及这一过程是如何在自然酶的作用下发生的。玛丽·施韦策还发现了恐龙和鸡之间的分子相似性，为二者之间的关联提供了进一步的证据。

图书在版编目（CIP）数据

冈特生态童书.第四辑：修订版：全36册：汉英对照 /
（比）冈特·鲍利著；（哥伦）凯瑟琳娜·巴赫绘；
何家振等译.—上海：上海远东出版社，2023
书名原文：Gunter's Fables
ISBN 978-7-5476-1931-5

Ⅰ.①冈… Ⅱ.①冈… ②凯… ③何… Ⅲ.①生态环
境–环境保护–儿童读物—汉、英 Ⅳ.①X171.1–49

中国国家版本馆CIP数据核字(2023)第120983号
著作权合同登记号图字09-2023-0612号

策　　划　张　蓉
责任编辑　曹　茜
封面设计　魏　来　李　廉

冈特生态童书

化装舞会

[比]冈特·鲍利　著
[哥伦]凯瑟琳娜·巴赫　绘
何家振　译

记得要和身边的小朋友分享环保知识哦！
八喜冰淇淋祝你成为环保小使者！